Light is Everywhere:
Sources of Light and Its Uses

(For Early Learners)

SPEEDY
PUBLISHING

Speedy Publishing LLC
40 E. Main St. #1156
Newark, DE 19711
www.speedypublishing.com

What would the world be like without light? Imagine if there were no light from the sun? No plants would grow because plants rely on light to make food. Animals and then humans too would have nothing to eat.

Sunlight is the source of most of the energy on Earth. Without the sun the Earth would become very cold and the water would freeze.

Natural Light

The sun is the most important source of light

The stars and moon are also natural forms of light

Artificial Light

Candles, fire and oil lamps

The electric light is now the most convenient form of artificial light

The light bulb, neon light and fluorescent tube are types of electric light

Lasers are also artificial lights.

Other facts of Light

Light is a form of energy.

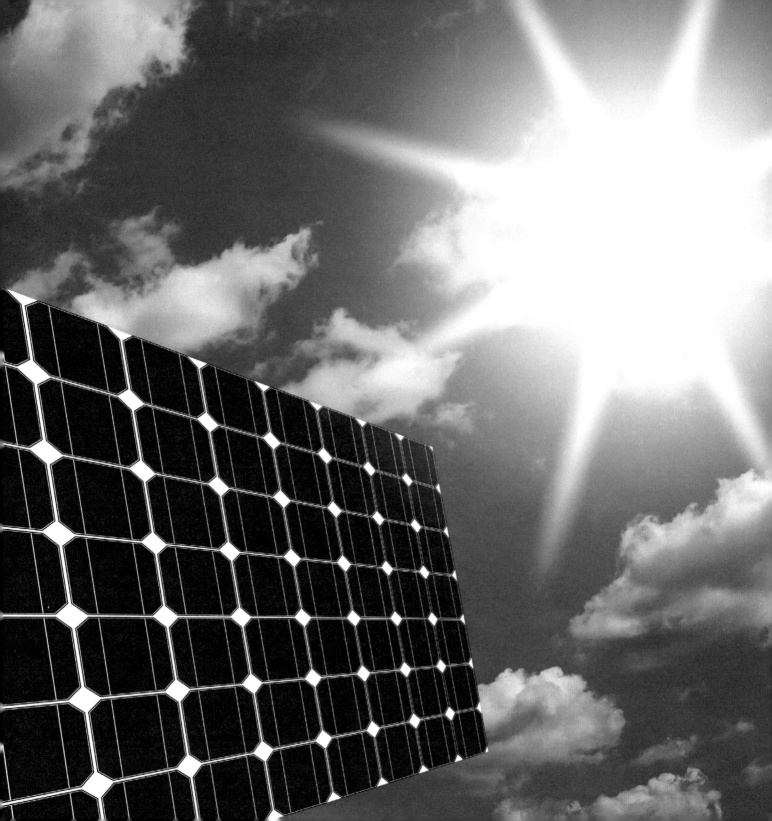

Light travels in waves.

The speed of light is the fastest.

Light waves usually travel in straight lines called rays of light.

Light waves may be reflected or refracted and may change speed depending on the material they are passing through.

Made in the USA
Las Vegas, NV
15 September 2021